GOODS OF THE MIND, LLC

Competitive Mathematics Series

for

Gifted Students in Grades 3 and 4

PRACTICE WORD PROBLEMS

Cleo Borac, M. Sc.
Silviu Borac, Ph. D.

This edition published in 2013 in the United States of America.

Editing and proofreading: David Borac, B.Mus.
Technical support: Andrei T. Borac, B.A., PBK

Send all inquiries to:

Goods of the Mind, LLC
1138 Grand Teton Dr.
Pacifica
CA, 94044

Competitive Mathematics Series for Gifted Students
Level II (Grades 3 and 4)
Word Problems

Contents

FOREWORD

The goal of these booklets is to provide a problem solving training ground starting from the earliest years of a student's mathematical development.

In our experience, we have found that teaching how to solve problems should focus not only on finding correct answers but also on finding better solution strategies. While the correct answer to a problem can typically be obtained in several different ways, not all these ways are equally useful for learning how to solve problems.

The most basic strategy is *brute force*. For example, if a problem asks for the number of ways Lila and Dina can sit on a bench, it is easy to write down all the possibilities: Dina, Lila and Lila, Dina. We arrive at this solution by performing all the possible actions allowed by the problem, leaving nothing to the imagination. For this last reason, this approach is called brute force.

Obviously, if we had to figure out the number of ways 30 people could stand in a line, then brute force would not be as practical, as it would take a prohibitively long time to apply.

Using brute force to obtain the correct answer for a simpler problem is not necessarily a useful learning experience for solving a similar problem that is more complex. Moreover, solving problems in a quantitative manner, assuming that the student can transfer simple strategies to similar but more complex problems, is not an efficient way of learning problem solving.

From this simple example, we see that the goal of *practicing* problem solving is different from the goal of problem solving. While the goal of problem solving is to obtain a correct answer, the goal of practicing problem solving is to acquire the ability to develop strategies, generate ideas, and combine approaches that are powerful enough to solve the problem at hand as well as future similar problems.

While brute force is not a useless strategy, it is not a key that opens every door. Nevertheless, there are problems where brute force can be a useful tool. For instance, brute force can be used as a first step in solving a complex problem: a smaller scale example can be approached using brute force to help the problem solver understand the mechanics of the problem and generate ideas for solving the larger case.

All too often, we encounter students who can quickly solve simple problems by applying brute force and who become frustrated when the solving methods they have been employing successfully for years become inefficient once problems increase in complexity. Often, neither the student nor the parent has a clear understanding of why the student has stagnated at a certain level. When the only arrows in the quiver are guess-and-check and brute force, the ability to take down larger game is limited.

Our series of books aims to address this tendency to continue on the beaten path - which usually generates so much praise for the gifted student in the early years of schooling - by offering a challenging set of questions meant to build up an understanding of the problem solving process. Solving problems should never be easy! To be useful, to represent actual training, problem solving should be challenging. There should always be a sense of difficulty, otherwise there is no elation upon finding the solution.

Indeed, practicing problem solving is important and useful only as a means of learning how to develop better strategies. We must constantly learn and invent new strategies while questioning the limitations of the strategies we are using. Obtaining the correct answer is only the natural outcome of having applied a strategy that worked for a particular problem in the time available to solve it. Obtaining the wrong answer is not necessarily a bad outcome; it provides insight into the fallacies of the method used or into the errors of execution that may have occured. As long as students manifest an interest in figuring out strategies, the process of problem solving should be rewarding in itself.

Sitting and thinking in a focused manner is difficult to train, particularly since the modern lifestyle is not conducive to adopting open-ended activities. This is why we would like to encourage parents to pull back from a quantitative approach to mathematical education based on repetition, number of completed pages, and the number of correct answers. Instead, open up the time boundaries that are dedicated to math, adopt math as a game played in the family, initiate a math dialogue, and let the student take his or her time to think up clever solutions.

Figuring out strategies is much more of a game than the mechanical repetition of stepwise problem solving recipes that textbooks so profusely provide, in order to "make math easy." Mathematics is not meant to be easy; it is meant to be interesting.

Solving a problem in different ways is a good way of comparing the merits of each method - another reason for not making the correct answer the primary goal of the activity. Which method is more labor intensive, takes more time or is more prone to execution errors? These are questions that must be part of the problem solving process.

In the end, it is not the quantity of problems solved, the level of theory absorbed, or the number of solutions offered in ready-made form by so many courses and camps, but the willingness to ask questions, understand and explore limitations, and derive new information from scratch, that are the cornerstones of a sound training for problem solvers.

These booklets are not a complete guide to the problem solving universe, but they are meant to help parents and educators work in the direction that, aside from being the most efficient, is the more interesting and rewarding one.

The series is designed for mathematically gifted students. Each book addresses an age range as some students will be ready for this content earlier, others later. If a topic seems too difficult, simply try it again in a couple of months.

Problems that involve comparing quantities are often solved using a concrete model. The typical method is to represent quantities by boxes of appropriate sizes.

Example 1 In Alfonso's shop, one lb of coffee and 3 lbs of lemons cost 14 dollars. One pound of coffee is 4 times more expensive than one pound of lemons. How much does a pound of coffee cost?

Solution

By weight, lemons are less expensive than coffee. Use a figure, such as a small rectangle, to represent the cost of one pound of lemons. Then represent the cost of one pound of coffee by a rectangle made up of 4 small rectangles.

Cost of:

1 lb lemons

1 lb coffee

Now, make a diagram for the combined cost of 3 lbs of lemons and one lb of coffee:

Cost of 1 lb coffee and 3 lb lemons:

= $ 14

According to the diagram, each small rectangle represents 2 dollars. Therefore, a pound of lemons costs 2 dollars and a pound of coffee costs 8 dollars.

Reduction to Unity is an elementary method used to account for direct and inverse relationships.

Example 1

40 walnut trees produce 3600 lbs of nuts over a 6 year span. How many walnut trees produce 2040 lbs of nuts over an 8 year span?

years	walnuts	production	description
6	40	3600	40 trees produce 3600 lbs in 6 years
1	40	600	40 trees produce 600 lbs in 1 year
1	1	15	1 tree produces 15 lbs in 1 year
8	1	120	1 tree produces 120 lbs in 8 years
8	17	2040	17 trees produce 2040 lbs in 8 years

Experiment

60 deer graze 540 lbs of grass over a 3 hour interval. How many deer will graze 300 lbs of grass over a 2 hour interval? Fill in the missing values:

deer	60	60	1	1	z
hours	3	1	1	y	2
lbs grass	540	x	3	6	300

Answer: $x = 180,$ $y = 2,$ $z = 50.$

Practice One

Do not use a calculator for any of the problems!

Exercise 1

The Wicked Witch of the West's hat is 6 inches long. Every time she casts an evil spell, the hat doubles in height. Every time she casts a good spell, the hat shrinks by 5 inches.

1. How tall is the hat after she casts 3 evil spells followed by 2 good spells?

2. How tall is the hat after she casts an evil spell, a good spell, an evil spell, a good spell, and an evil spell (in that order)?

Exercise 2

Dina wanted to buy some flowers for Mother's Day. She bought some beautiful flowers and a nice vase. She spent 44 dollars in total. The flowers cost 10 dollars more than the vase. How much did the flowers cost?

Exercise 3

Dina had 14 glo cards and Lila had 19 glo cards. They decided to split these evenly with Amira. How many glo cards did Lila give Amira?

Exercise 4

Stephan and Amanda were traveling to Fiji. They packed a lot of things and exceeded the baggage allowance by 50 lbs. Together, they paid 200 dollars for excess weight: Stephan paid 80 and Amanda paid the rest. How many pounds heavier was Amanda's luggage than Stephan's?

Exercise 5

Lila's mother is 28 years older than Lila and 31 years older than Amira. Five years from now, how much older than Amira will Lila be?

Exercise 6

Max, the baker, is sending some of his breads to a local restaurant. Of the 96 loaves he is sending, half are baguettes, a third are Vienna breads, and the rest are boules. How many more baguettes are there than boules?

Exercise 7

King Minos had three large holding tanks for storing wheat grain. Each tank had a capacity of 1200 liters. In June, the Minoans stored 1000 liters of grain. In July, they stored 200 liters more. At most how many more liters than in June would they have been able to store in August?

Exercise 8

Amira is trading toys with Lila. Amira traded 20 plastic animals for a doll. Then, she got tired of the doll and exchanged her for 30 of Lila's animals. Lila lost one of the doll's shoes and traded her back to Amira for only 10 animals. How many animals did the doll cost Amira in the end?

Exercise 9

Arbax, the Dalmatian, and Lynda, the terrier, found a bag of dog treats. Completely forgetting his manners, Arbax quickly ate a third of them. Lynda pushed him aside with unladylike vigor and ate half of the remaining treats. Then they shared the leftover treats - 5 treats each. How many treats did Arbax eat in total?

Exercise 10

Max, the baker, had two identical boxes of bagels: one full and the other only two thirds full. The boxes weighed 10 lbs and 7 lbs. How many pounds of bagels did Max have to move from one box to the other so that they both weighed the same?

Exercise 11

There were 10 groups of penguins on a beach. There were three penguins in the smallest group: Alpha, Kappa, and Omicron. Zeta came along and ordered the penguins in each group except the smallest group to send two penguins to each of the other groups. In this way, she reckoned, each group would have the same number of penguins. How many penguins were there in total?

Exercise 12

Dina went to Max's bakery to buy some treats. She bought twice as many strudels as croissants. On the way home, Arbax grabbed the bag of strudels and ate 3 of them before Dina could stop him. Dina decided to go back to the store to replace them but, by the time she got there, the strudels had sold out. Dina decided to purchase 6 more croissants. Now she had 3 times as many croissants as strudels. How many pieces of pastry did Dina take home?

Exercise 13

Four penguins swim around an island. There are four landmarks around the island arranged in clockwise order: the Stream, the Big Rock, the Wreck, and the Icemaker. There are 80 miles of swimming from the Stream to the Big Rock, 120 miles from the Big Rock to the Wreck, 130 miles from the Wreck to the Icemaker, and 70 miles from the Icemaker to the Stream. The distance around the island from the Stream to the Wreck (via the Big Rock) is the same as from the Wreck to the Stream (via the Icemaker). What is this distance?

200 mi

Exercise 14

Max, the baker, must divide 60 lbs of sugar into two bags so that one bag has 20 lbs less sugar in it than the other. How heavy will each bag be when he has finished?

40 lbs and 20 lbs

Exercise 15

Amira celebrated her 5th birthday in style. She invited all her friends and they all sat around the big garden table. Amira placed 18 chairs at equal distances around the table and labeled them from 1 to 18. Dina sat on chair number 14 and Lila sat opposite from Dina. What was the label on Lila's chair?

Exercise 16

15 scouts eat 1050 bagels in 35 days. How many scouts eat 360 bagels in 18 days?

20

MULTIPLES AND FRACTIONS

1. A non-zero integer has an infinite number of multiples. For example, the multiples of 7 form the infinite sequence:

   $$\ldots, -49, -35, -28, -21, -14, -7, 0, 7, 14, 21, 28, 35, 49, \ldots$$

2. All infinite sequences of multiples include the number 0. Zero is a multiple of any integer.

3. A multiple of a number is said to be *divisible* by that number. For example, 14 is divisible by 7. Of course, zero is also divisible by 7.

4. Conversely, we say that 7 is a *divisor* of 14.

5. The number of divisors of an integer is finite. For example. the divisors of 6 are 1, 2, 3, and 6.

6. 1 is a divisor of any integer.

7. Zero is not a divisor of any number. Division by zero is not defined.

PRACTICE TWO

Exercise 1

Lila chose an integer number. Then she told Dina: "If I add half of my number to a third of it, I get 500. Can you guess my number?" What was Dina's reply?

600

Exercise 2

The sum of two numbers is 791. The sum of the triple of one of the numbers and the other number is 875. What is the positive difference of the numbers?

Exercise 3

Stephan, the tennis coach, has 32 tennis balls in a bag. Of these, there are four times as many Dunlap balls as Wilson balls. Which of the following statements must be true?

(A) There are 8 Wilson balls.

(B) Stephan must have balls of at least one more brand in the bag.

(C) If you take out a ball without looking, it will be a Dunlap.

(D) Wilson brand is not Stephan's favorite.

(E) The bag is full of balls.

Exercise 4

Dina, Lila, and Amira went to pick apricots. Dina picked as many as Lila and Amira combined. Lila picked one fifth the amount Amira picked. Which of the following could be the total number of apricots they picked?

(A) 91

(B) 117

(C) 132

(D) 143

(E) 154

Exercise 5

Lila said to Dina: "Let's play with your mini-monsters! You have 94, right?" Dina replied: "I don't have 94 mini-monsters. If I had one third of what I have plus another sixth of what I have, plus the double of what I have, and if I included the 4 farm animals, then I would have 94 toys." How many mini-monsters did Dina actually have?

Exercise 6

Arbax, the Dalmatian, is organizing his bone collection. He finds that half of the bones are way too long and breaks them in half. Then, he finds that 36 of the newly broken bones are still too long and breaks those in half too. In the end, he has 96 pieces to store. How many bones did he start with?

Exercise 7

Max, the baker, has been selling bags of 10 pretzels for 5 dollars a piece all day. Now it is evening and Max wants to get rid of as much of his merchandise as he can before it becomes day old. He decides to keep the price of each bag the same but add some more pretzels to each bag so that the final price is 4 dollars per dozen. How many pretzels did Max add to each bag?

Exercise 8

Stephan is moving to another tennis club and is now weighing his boxes to get a moving quote. He has two identical boxes: one with 55 tennis balls, and the other with 15 tennis balls. The box with 55 balls weighs 3 kg and the box with 15 balls weighs 1 kilogram. How much does an empty box weigh?

Exercise 9

Divide the number 29 into three integer parts so that, if we add 1 to the first part, 2 to the second part, and 4 to the third part, you get consecutive numbers.

10, 10, and 9

Exercise 10

Divide the number 72 into four integer parts so that if we add 5 to the first part, subtract 5 from the second part, multiply the third part by 5, and divide the fourth part by 5, all the results are the same.

Exercise 11

Amira's teacher told het to multiply a number by 7 and subtract 5. Amira absent-mindedly multiplied the number by 5 and added 7. However, Amira's answer was correct! What number did the teacher give her?

Exercise 12

In a game of Ask-Me-NOT, a player receives 1 point for each correct answer and loses 1 point for each incorrect answer. Stephan was invited to such a game and, overall, was able to give 5 correct answers. Nevertheless, he ended the game with a score of −12. How many questions did he answer in total?

17 questions

Exercise 13

Jared works at a pet store. His job is to sort crickets and worms into lunch sized portions for the lizards and birds that live in the store. He puts either 5 crickets or 7 worms in each box. If he has a total of 46 worms and crickets, how many boxes will he fill?

8 boxes

Exercise 14

In Dina's chicken coop, there are 4 red chickens for every 5 white chickens. The total number of chickens could be:

(A) 61

(B) 63 .

(C) 75

(D) 100

(E) 101

Exercise 15

There are 14 dolphins playing around Amanda's boat. 12 of them are females and 8 of them are calves. What is the smallest possible number of female calves? What is the largest possible number?

Exercise 16

Ali's donkey is so stubborn that Ali has to drag him forward. The donkey will follow Ali for 20 feet, after which he walks backwards for 5 feet and waits for Ali to drag him again. After much struggle, Ali was able to move the donkey 100 feet, from the Fountain to the Market. How many feet did the donkey walk in total?

125 Feet

Exercise 17

Dina, Lila, and Amira are cleaning out their toy collections. Lila had 10 dinosaurs more than Dina and 10 dinosaurs fewer than Amira. They sold the dinosaurs they no longer wanted at a yard sale. Dina sold a quarter of her dinos, Lila sold half of hers, and Amira sold 14. They were left with a total of 47 dinos. How many dinos did Dina have in the beginning?

Exercise 18

Arbax sorted his bones in piles of 6 bones each. After making the piles, he noticed that the number of bones left over was equal to half the number of piles. The total number of bones cannot be:

(A) 13

(B) 14

(C) 26

(D) 65

(E) 52

Exercise 19

Max baked 38 baguettes and put them on display to the left and to the right of his sales counter. Lila comes in with some friends and they each make an observation about the number of baguettes on the left and the number of baguettes on the right:

(A) there is an equal number of baguettes on both sides

(B) there are three times more baguettes on the left

(C) there are six baguettes more on the right

(D) there are four baguettes fewer on the left

(E) there are ten baguettes more on the right

Which observation cannot be correct?

Exercise 20

At the Lucky Stars merry-go-round, Lila passes through a tunnel, by a mirror, and over a bottomless pit. She collects 3 points at the tunnel, 4 points at the mirror, and 5 points at the bottomless pit. During the ride, she collected a total of 100 points. How many times did she pass by the mirror?

Exercise 21

Arbax thinks to himself: "If I had 8 bones more than I have right now, then I could share them equally with Lynda and still have one bone more than I have now." How many bones does Arbax have?

Exercise 22

Max is closing his bakery at the end of the day. He thinks to himself: "If I had sold three times as many baguettes as focaccias or if I had sold twice as many foccacias as baguettes, I would have sold 50 breads more than I did." How many breads did Max sell that day?

Exercise 23

Stephan, the tennis coach, is visiting a new client. This client lives on the same street as another client and Stephan has forgotten which one of them lives at house number 38 and which one at 78. He calls the new client and explains his problem. It is, however, April 1st and Stephan gets the following riddle instead of direct help: "If you multiply my house number by 15 and multiply the other house number by 14, the sum of the two results will be a multiple of three." Help Stephan quickly!

Exercise 24

Which of the following is the correct answer to the question: "How many numbers from 0 to 40 are multiples of 7?"

(A) $\{7, 14, 21, 28, 35\}$

(B) 5 numbers

(C) 6 numbers

(D) only 1 and 7 since 7 is prime

REMAINDERS

Students must understand the difference between integer division and long division with decimals. The concepts of *multiple, divisor,* and *remainder* are specific to integer arithmetic. For problems that involve *integer division*, the following theorem must be used:

For any two integers D and d, with d non-zero, there exist two uniquely determined integers q and r, with $0 \le r < q$, such that:

$$D = d \times q + r$$

Any integer division must be put in this form - which does not involve dividing! Use only multiplication and addition.

The important facts to keep in mind when solving problems are:

1. The divisor d cannot be zero.
2. D can be zero, in which case q and r are also zero while d can have any value, as in: $0 = d \times 0 + 0$.
3. The remainder r ranges from 0 (when D is divisible by d) to $d-1$.
4. If the remainder is zero, d is a *divisor (factor)* of D.
5. If the remainder is zero, D is a *multiple* of d.
6. Zero is a multiple of any integer.
7. Zero is not a divisor (factor) of any integer.
8. An integer has an infinite number of multiples.
9. An integer has a finite number of factors: 1 has only one factor, prime numbers have 2 factors, and composite numbers have various numbers of factors.

PRACTICE THREE

Exercise 1

In a bag of coins, there are quarters, dimes, and nickels. Nyame asks Anansi to exchange as many one dollar bills as he can into quarters. Then, Nyame asks Turtle to exchange as many one dollar bills as he can into dimes. Finally, Nyame asks Firefly to exchange as many one dollar bills as he can into nickels. What is the largest possible dollar amount of the coins left in the bag?

Exercise 2

If Amira had one doll more than she has, she could arrange her dolls in groups of five exactly. If Dina had two dolls more than she has, she could arrange her dolls in groups of five exactly. If Lila had three dolls less than she has, she could arrange them in groups of five exactly. If the three girls put all their dolls together, would it be possible to arrange them in groups of five without any toys left over?

Exercise 3

Lila was making loot bags for her birthday. She had 8 loot bags and wanted to put chocolates from a larger box in each of them. She placed one chocolate in each bag and continued to do so until each bag contained 13 chocolates. When she had finished, there were some chocolates left in the larger box, but not enough to place one more in each loot bag. At the start, the large box may have contained:

(A) 100 chocolates.

(B) 108 chocolates.

(C) 112 chocolates.

(D) 134 chocolates.

Exercise 4

Dina and Lila divided 122 by 17. Dina got a quotient of 7 and a remainder of 3. Lila got a quotient of 6 and a remainder of 20. Which of the following is true? (check all that apply)

(A) only Dina is correct

(B) only Lila is correct

(C) both Lila and Dina are correct

(D) neither Lila nor Dina are correct

Exercise 5

Max, the baker, bought several 8 lbs bags of flour. He purchased 136 lbs of flour in total and used a quarter of this quantity the next day. How many bags were left unopened?

Exercise 6

Stephan, the tennis coach, ordered some tennis balls. He received 10 boxes, each containing 12 cylinders of 3 balls, and some boxes of loose balls, each containing 12 balls. If he received 408 balls in total, how many boxes of loose balls was he sent?

Exercise 7

Dina had 12 dinosaurs, Lila had 20 dinosaurs, and Amira had more than 10 dinosaurs. Dina, Lila, and Amira counted all their dinosaurs and found out that they could divide them equally. What is the smallest number of dinosaurs Amira could have had?

Exercise 8

Arbax has dug up his bone collection. He made piles of 5 bones each and gave any leftover bones to Lynda. Then he made piles of 4 bones each and again gave any leftover bones to Lynda. Lynda cannot have:

(A) 4 bones

(B) 5 bones

(C) 6 bones

(D) 7 bones

(E) 8 bones

Exercise 9

Alfonso, the grocer, is filling some gift baskets with apples and mangoes. He does not want to count all the fruit but he notices that, if he tries to put an equal number of apples in each basket, there are 4 apples in each basket and 8 apples left over. If he tries to put an equal number of mangoes in each basket, there are 5 mangoes in each basket and 5 mangoes left over. If there are 7 more mangoes than apples, can you find out how many baskets he has?

Exercise 10

Dina and Lila have set up 61 decorations for Hallowe'en. After they finished, Lila said to Dina: "I put up 5 more decorations than a third of the number of decorations you put up." How many decorations did Dina put up?

Exercise 11

Max, the baker, has just finished baking his daily supply of bread. He notices there are 4 baguettes for each boule and 6 rolls for each baguette. The total number of baguettes, boules, and rolls is 290. How many are there of each?

Exercise 12

Lila, Dina, and Amira helped Stephan harvest the pomegranates in his tree. Each girl put the pomegranates she harvested in boxes of 8 and tossed the leftover fruit in a basket. At the end of the harvest, what was the largest number of boxes that could be filled from the contents of the basket?

TIME, RATES, AND COINS

Military Time is often used in problem statements. Military time is also the way people express time in many countries around the world, most notably in Europe.

1. Military time uses a 24 hour day instead of two 12 hour periods designated as AM (ante meridiem) and PM (post meridiem).

2. In military time, the hour counter goes from 00 to 23.

3. In military time, the minute and second counters go from 00 to 59.

Example

How many times in 24 hours does a clock set to show military time display exactly three digits of 3 simultaneously?

Answer Three times: $03:33$, $13:33$, and $23:33$.

How many times in 24 hours does a clock set to show military time display exactly three digits of 4 simultaneously?

Answer Two times: $04:44$ and $14:44$.

How many times in 24 hours does a clock set to show military time display exactly three digits of 5 simultaneously?

Answer Two times: $05:55$ and $15:55$.

A **rate** shows how a quantity changes over time. In problems at this level, we assume that all rates are constant. This means that objects never "speed up" or "slow down." Instead, objects always travel at the same speed. Unless otherwise specified, objects are considered to be *pointlike*. This means they have no size of their own.

Since problems with rates handle *physical quantities and not just numbers*, we have to make sure we compare them using the same unit! In problems with rates, it is sometimes necessary to *change the units* to make them all the same. Before attempting the problems in this section, students should review *metric system conversions*.

For coin problems, remember that the *number of coins* is different from the *monetary value* of the coins.

Example

In a piggy bank there are 4 times as many nickels as quarters and as many quarters as dimes. The total amount in the piggy bank is 6.50 dollars. How many quarters are there in the piggy bank?

Answer For *each quarter*, there are two dimes and four nickels. As monetary value, we can say that *for every* 25 *cents, there are another* $1 \times 10 + 4 \times 5 = 40$ cents in the piggy bank. Therefore, the amount in the piggy bank can be divided into groups of coins that are worth 65 cents. Since the total amount of money in the bank is 6.50 dollars, there must be 10 such groups. Each group contains one quarter, so there must be 10 quarters in the piggy bank.

PRACTICE FOUR

Do not use a calculator for any of the problems!

Exercise 1

Four clocks show the following times:

(A) 4:00 PM

(B) 4:30 PM

(C) 4:45 PM

(D) 4:50 PM

One of them stopped 15 minutes before Dina left for school. Another stopped one third of an hour after Dina arrived at school. The other two clocks are broken and show random times. If it takes Dina 10 minutes to walk from home to school, which clocks show random times?

Exercise 2

Lila left for her singing class at 4:20 PM and returned at 7:10 PM. For how many minutes was Lila away from home?

Exercise 3

Dina's mother is four times older than Dina. Dina is 35 years younger than her father. Dina's father was 5 years older than Dina's mother when Dina was born. How old is Dina now?

Exercise 4

Amalia, the dance instructor, has started her summer classes. On weekdays, she teaches from 9:00 AM to 11:30 AM and from 1:00 PM to 5:00 PM. On weekends, she teaches from 1:00 pm to 4:00 pm. If all her classes are the same length, each class cannot be:

(A) 10 minutes long.

(B) 15 minutes long.

(C) 20 minutes long.

(D) 12 minutes long.

Exercise 5

Dina and Lila each recieved an hourglass. The two hourglasses are not the same size. Lila has to turn her hourglass 4 times in 5 hours and Dina has to turn her hourglass 6 times in 5 hours. What is the difference between the time intervals measured by the two hourglasses?

Exercise 6

Express the following military times as AM or PM.

1. 06:00
2. 12:30
3. 23:45
4. 19:12
5. 02:30
6. 14:15
7. 11:30

Exercise 7

Alfonso has a clock in his grocery store that puzzles all his clients. This is what the clock shows at various times during the day:

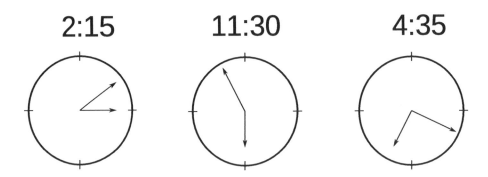

2:15   11:30   4:35

One day, Dina finally figured out what was wrong with the clock. She told Lila: "When it is 9:20, this clock shows .... " Lila could not hear the rest because a firetruck passed by. Can you finish Dina's sentence?

Exercise 8

Amira broke her piggy bank to buy herself a movie ticket. After she counted her savings, she exclaimed: "If I had 20 dimes more, 5 nickels less, and 5 pennies less, then I would have six times the amount I have now!" How much money, in dollars, did Amira find in the piggy bank?

Exercise 9

Amira's mother was in her garden, tying tomato plants to wooden stakes: it took her 10 minutes to stake each plant and she placed the plants 2 feet apart. If she started at one end of the row at 8:00 AM, how far from the first plant was the tomato plant she finished tying up at 9:00 AM?

Exercise 10

Write the next term in the sequence:

022814, 033115, 043016, 053117, ...

Exercise 11

Dina rides on horseback from one end of a trail while Lila rides on horseback from the other end of the trail. Dina rides mostly uphill, so her horse moves at an average of 2 miles per hour. Lila rides mostly downhill. Her horse averages 3 miles per hour. Lila and Dina meet after 90 minutes of riding. How long is the trail?

Exercise 12

Dina and Lila went cross country skiing. They left from different ends of the same trail and decided to meet at a lodge which was midway. Dina started 3 miles from the lodge and Lila started 5 miles from the lodge. If they both skied at the same average speed of 2 miles per hour and started at the same time, how long did Dina have to wait for Lila at the lodge?

Exercise 13

Dina's mother can clean the floor of the house in 3 hours. Dina's father can do the same job in 2 hours. How long does it take them to clean the floor if they work together?

Exercise 14

Dina's mother has purchased a robot that can clean the floor of the house in 2 hours. Dina's father did not know about this, so he also purchased the same model of robot. How long does it take the two robots to clean the floor if they work at the same time?

Exercise 15

Max left for Carapatas and drove at an average speed of 80 miles per hour. After 2 hours, Max got a traffic ticket for speeding. Humbled, he continued at an average speed of 70 miles per hour. The distance from Max's home to Carapatas is 440 miles. How long did it take Max to get from his home to Carapatas if his encounter with the traffic police took 15 minutes?

Exercise 16

Amira bikes on the beach trail from the parking lot to the kayak store. Dina starts biking from the parking lot 10 minutes after Amira started and arrives at the kayak store 10 minutes before Amira. If Dina rides twice as fast as Amira, how long did Amira ride?

Exercise 17

Complete the missing values:

minutes	30	20	45				24		
fraction of one hour	$\frac{1}{2}$			$\frac{1}{5}$	$\frac{1}{10}$	$\frac{3}{10}$		$\frac{4}{5}$	$\frac{5}{12}$

Exercise 18

Amalia, the dance instructor, has been invited to dinner by Alfonso, the grocer. Amalia figured she needed 45 minutes to get from her studio to the restaurant. Not wanting to be late, she left 15 minutes early. There was some roadwork on the way and traffic was reduced to one lane. She arrived at the restaurant at 6:20 PM. Alfonso was already there and exclaimed: "You are only 10 minutes late!" When should Amalia have left to be sure that she arrived on time?

Exercise 19

As Dina walked on stage for her piano recital, she noticed that the time shown in military notation had the largest digit sum possible. What time did Dina walk on stage?

Exercise 20

Amira's cousin was born on the day in the 1990's when the date (in the form month/day/year) had the largest possible digit sum. What was the exact date of his birth?

MISCELLANEOUS PRACTICE

Do not use a calculator for any of the problems!

Exercise 1

At a party, there are only groups of three siblings. As it happens, there is no group with only girls or only boys. Moreover, there are more than 2 groups in total and the number of boys equals the number of girls. The total number of girls can be:

(A) any multiple of 3

(B) an odd multiple of 3

(C) an even multiple of 3

(D) any multiple of 3 larger than 3

Exercise 2

Alfonso's house is at number 101 on Grapevine Street. There are five more houses from there until the end of the street. How many houses are there on Alfonso's side of the street (House numbers on the street start at 1 and they are odd on one side, even on the other)?

Exercise 3

Stephan, the tennis coach, set up a mock tournament for 4 of his students. Every time a student played another student, Stephan gave 5 dollars to the winner. If each student played all the other students, how much money did Stephan pay out in prizes?

Exercise 4

At running, Amira is the worst of the first 10 students in her class as well as the best of the last 10 students in her class. How many students are there in Amira's class?

Exercise 5

Julien, the Parisian master baker, only taught his secret French baguette recipe to one of his apprentices: Max. Max did not keep the recipe a secret and taught it to four of his apprentices. These apprentices went on to teach it to four apprentices each, who all taught the recipe to four more apprentices each as well. How many bakers know Julien's recipe now?

Exercise 6

Alfonso, the grocer, gave Lila a job over the summer: to canvas the neighborhood and take orders for deliveries of fresh vegetables. Lila had a sales quota of 60 dollars per day, but her neighbors trusted Alfonso so much that she managed to take orders of 80 dollars each day. At this pace, she fulfilled the total sales quota 5 days before her job officially ended. For how many days was her job supposed to last?

Exercise 7

Dina takes Russian language lessons each Monday, Wednesday, and Friday. What is the largest possible number of lessons Dina could take in any one month?

Exercise 8

Max, the baker, is placing fresh loaves of bread on the display shelf of his bakery. On the topmost shelf there are some large loaves of bread. On the middle shelf, there are three times more loaves than on the top shelf. On the bottom shelf, there are three times more loaves than on the middle shelf. The loaves on display are only one third of the number of loaves that Max baked that morning. In total, Max baked 585 loaves. How many loaves are there on the middle shelf?

Exercise 9

Stephan organized a local tennis tournament. His suppliers sent 4 boxes of tennis balls. In each box, there were 10 bundles and in each bundle there were 12 cylinders of three balls. On the first day of the tournament, Stephan placed one box in each of the two courts. The players in one of the cours used 43 balls, while the players in the other court used 53 balls. At the end of the day, how many tennis balls were still in unopened cylinders?

Exercise 10

Jared works at a pet store and his duties include separating the daily shipment of crickets into boxes. Today, the shipment contained between 110 and 150 crickets, but no one at the store knew exactly how many. Jared tried to put 5 crickets in each box, but one cricket was left over. Then, he tried to put 6 crickets in each box, but once again, one cricket was left over. How many crickets were there in the shipment?

Exercise 11

Jared has been given an even harder problem to solve. He has to package worms and roaches! If he tries to make packages with one worm and one roach in each, there are 15 roaches left over. If he tries to make packages with 2 roaches and one worm in each, there are 17 worms left over. How many roaches and worms does he have to package?

Exercise 12

In Amira's class, there are 20 students. Their teacher gives them points for all sorts of wonderful reasons. As it happens, each student has a different number of points. The student with the most points has 100 points. The teacher says: "I have noticed that, if each of you would give one point to every student who have less points than you do, then you would all have the same number of points." How many points does the student with the smallest number of points have?

Exercise 13

Dina's mother purchased vegetables from Alfonso's shop. A pound of red peppers cost twice as much as a pound of sweet potatoes. By weight, Dina's mother put twice as large a quantity of sweet potatoes as of red peppers in a bag. Alfonso told her she had to pay 20 dollars but Dina's mother did not have enough money in her purse. Alfonso said: "If you purchase only half of the sweet potatoes, the total will be exactly as much as you have." How much money did she have?

Exercise 14

On Monday morning at 8:00 AM, Jared started to clean the 28 fish tanks in the pet store. It took Jared 25 minutes to clean each of the first 8 tanks. After that, he made some improvements to his process and cut 5 minutes from the cleaning time per tank for all of the remaining tanks. If Jared spent 4 hours per day cleaning the tanks, what day of the week was it when he finished?

Exercise 15

In a singles tennis tournament, pairs of opponents are selected at random for each round. The winners qualify for the next round. After four rounds, there were 4 players in the semi-finals. How many players participated in total?

Exercise 16

A basket with 10 potatoes in it weighs 5 lbs. If Alfonso adds another 10 potatoes, it weighs 7 lbs. Alfonso wants to figure out how much the empty basket weighs. Can you help him?

Exercise 17

In 5 hours, 20 gophers dig 50 feet of gallery.

1. How many feet of gallery do 24 gophers dig in 3 hours?
2. How many hours does it take 12 gophers to dig 60 feet of gallery?
3. How many gophers will it take to dig 30 feet of gallery in 2 hours?

Exercise 18

Amira wrote lyrics for a song: "3 hours of homework, 2 hours of piano, are worth 8 hours in the sun. 2 hours of homework, 4 hours of piano, are worth 8 hours in the sun, eiiiiiight hours in the sun!" Now, she is trying to figure out how many hours in the sun one hour of piano is worth. Can you help her?

Exercise 19

Together, two taps fill an empty cistern with water in 4 hours. One of the taps can fill the cistern in 12 hours. How many hours does it take the other tap to fill the cistern?

Exercise 20

A farmer has 4 horses, 2 cows, and 20 sheep. A cow eats three times as much as a sheep and a horse eats four times as much as a sheep. The farmer purchased 693 rations of feed, enough to last them 33 days. (The farmer purchases feed for a month but he likes to have a bit extra just in case.) After one week, however, the farmer took in his neighbor's 10 sheep as his neighbor was going on vacation to Hawaii. How many days earlier than expected did he run out of feed?

Exercise 21

Each group of four numbers in the sequence follows the same pattern. Which number is hidden by the question mark?

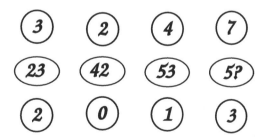

SOLUTIONS TO PRACTICE ONE

Exercise 1

The Wicked Witch of the West's hat is 6 inches long. Every time she casts an evil spell, the hat doubles in height. Every time she casts a good spell, the hat shrinks by 5 inches.

1. How tall is the hat after she casts 3 evil spells followed by 2 good spells?

2. How tall is the hat after she casts an evil spell, a good spell, an evil spell, a good spell, and an evil spell (in that order)?

Solution 1

1. Three bad spells increase the height of the hat to $6 \times 2 \times 2 \times 2 = 48$ inches and two good spells decrease the height by 10 inches. The hat is now $48 - 10 = 38$ inches tall.

2. The first evil spell increases the height of the hat to 12 inches. The good spell that follows decreases the height of the hat to 7 inches. The next evil spell increases the height to 14 inches and the following good spell decreases it to 9 inches. The final evil spell increases the height to 18 inches.

Exercise 2

Dina wanted to buy some flowers for Mother's Day. She bought some beautiful flowers and a nice vase. She spent 44 dollars in total. The flowers cost 10 dollars more than the vase. How much did the flowers cost?

Solution 2

If we add 10 dollars to the cost of the vase, it will cost as much as the flowers. Let us add 10 dollars to the cost of the vase plus the flowers, to get twice the price of the flowers: $44 + 10 = 54$. The flowers alone cost 27 dollars.

Exercise 3

Dina had 14 glo cards and Lila had 19 glo cards. They decided to split these evenly with Amira. How many glo cards did Lila give Amira?

Solution 3

Dina and Lila have $14 + 19 = 33$ glo cards in total. When the glo cards are split evenly, each of the three girls has 11 glo cards. Therefore, Amira received 8 glo cards from Lila and 3 glo cards from Dina.

Exercise 4

Stephan and Amanda were traveling to Fiji. They packed a lot of things and exceeded the baggage allowance by 50 lbs. Together, they paid 200 dollars for excess weight: Stephan paid 80 and Amanda paid the rest. How many pounds heavier was Amanda's luggage than Stephan's?

Solution 4

The luggage allowance is the same for every passenger, so the difference in weight comes only from excess luggage.

Stephan paid two fifths of the cost and Amanda paid three fifths. Therefore, in weight, Amanda's luggage exceeded Stephan's luggage by one fifth. Since the total excess weight was 50 lbs, Amanda's luggage is 10 lbs heavier than Stephan's.

Exercise 5

Lila's mother is 28 years older than Lila and 31 years older than Amira. Five years from now, how much older than Amira will Lila be?

Solution 5

Lila is $31 - 28 = 3$ years older than Amira. She will be 3 years older than Amira throughout their lives, including 5 years from now.

Exercise 6

Max, the baker, is sending some of his breads to a local restaurant. Of the 96 loaves he is sending, half are baguettes, a third are Vienna breads, and the rest are boules. How many more baguettes are there than boules?

Solution 6

The number of baguettes is: $96 \div 2 = 48$.

The number of Vienna breads is: $96 \div 3 = 32$.

The number of boules is: $96 - 48 - 32 = 16$.

There are $48 - 16 = 32$ more baguettes than boules.

Exercise 7

King Minos had three large holding tanks for storing wheat grain. Each tank had a capacity of 1200 liters. In June, the Minoans stored 1000 liters of grain. In July, they stored 200 liters more. At most how many more liters than in June would they have been able to store in August?

Solution 7

The total storage capacity is $3 \times 1200 = 3600$ liters.

In June, 1000 liters were stored.

In July, 1200 liters were stored.

As a result, the remaining capacity is at most $3600 - 1000 - 1200 = 1400$ liters.

Exercise 8

Amira is trading toys with Lila. Amira traded 20 plastic animals for a doll. Then, she got tired of the doll and exchanged her for 30 of Lila's animals. Lila lost one of the doll's shoes and traded her back to Amira for only 10 animals. How many animals did the doll cost Amira in the end?

Solution 8

Amira paid 20 animals once and then again 10 animals. She also gained

30 animals from one of the trades. Therefore, Amira did not lose or gain any animals. She got the doll for free!

Exercise 9

Arbax, the Dalmatian, and Lynda, the terrier, found a bag of dog treats. Completely forgetting his manners, Arbax quickly ate a third of them. Lynda pushed him aside with unladylike vigor and ate half of the remaining treats. Then they shared the leftover treats - 5 treats each. How many treats did Arbax eat in total?

Solution 9

Work backwards.

If there were 10 treats left in the bag for Arbax and Lynda to share, then this amount must equal what Lynda ate just before, since she had eaten half of the remaining treats.

Therefore, Lynda must have eaten 10 treats in the second step and there were 20 treats in total when Arbax finished eating the first third.

If 20 treats were left over after Arbax ate a third, then the 20 treats represent two thirds of the contents of the bag. The bag contained 30 treats to start with.

Arbax ate 15 treats.

Lynda also ate 15 treats.

Magically, they shared the treats equally!

Exercise 10

Max, the baker, had two identical boxes of bagels: one full and the other only two thirds full. The boxes weighed 10 lbs and 7 lbs. How many pounds of bagels did Max have to move from one box to the other so that they both weighed the same?

Solution 10

The difference in weight between the two boxes comes from the fact that the full box contains more bagels. The box itself weighs the same in both cases. The difference in bagels equals one third of the total amount of bagels in the full box.

By subtracting 7 from 10 we get the weight of a third of the bagels in the full box.

Since a third of the bagels in the full box weighs 3 lbs, the total amount of bagels must weigh 9 lbs and the box itself must weigh 1 lb.

There are 9 lbs of bagels in a box and 6 lbs in the other. To equalize the weights, Max must transfer 1.5 lbs of bagels from one box to the other.

Exercise 11

There were 10 groups of penguins on a beach. There were three penguins in the smallest group: Alpha, Kappa, and Omicron. Zeta came along and ordered the penguins in each group except the smallest group to send two penguins to each of the other groups. In this way, she reckoned, each group would have the same number of penguins. How many penguins were there in total?

Solution 11

Two penguins will come from each of the 9 larger groups to the smallest group, bringing the total number of penguins in this group to $9 \times 2 + 3 = 21$.

Since, at the end of these maneuvers, Omicron's group has 21 members and all ten groups are equal in size, the total number of penguins is 210.

Exercise 12

Dina went to Max's bakery to buy some treats. She bought twice as many strudels as croissants. On the way home, Arbax grabbed the bag of strudels and ate 3 of them before Dina could stop him. Dina decided to go back to the store to replace them but, by the time she got there, the strudels had sold out. Dina decided to purchase 6 more croissants. Now she had 3 times as many croissants as strudels. How many pieces of pastry did Dina take home?

Solution 12 The following diagram is not to scale.

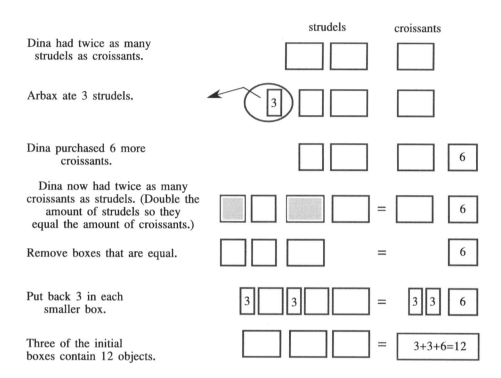

According to the diagram, Dina started with 8 strudels and 4 croissants.

She ended up with 5 strudels and 10 croissants and took home 15 pieces of pastry.

Exercise 13

Four penguins swim around an island. There are four landmarks around the island arranged in clockwise order: the Stream, the Big Rock, the Wreck, and the Icemaker. There are 80 miles of swimming from the Stream to the Big Rock, 120 miles from the Big Rock to the Wreck, 130 miles from the Wreck to the Icemaker, and 70 miles from the Icemaker to the Stream. The distance around the island from the Stream to the Wreck (via the Big Rock) is the same as from the Wreck to the Stream (via the Icemaker). What is this distance?

Solution 13

By adding all the distances together we get twice the distance required: $80 + 120 + 130 + 70 = 400$. The required distance is 200 miles.

Exercise 14

Max, the baker, must divide 60 lbs of sugar into two bags so that one bag has 20 lbs less sugar in it than the other. How heavy will each bag be when he has finished?

Solution 14

20 and 40.

Exercise 15

Amira celebrated her 5th birthday in style. She invited all her friends and they all sat around the big garden table. Amira placed 18 chairs at equal distances around the table and labeled them from 1 to 18. Dina sat on chair number 14 and Lila sat opposite from Dina. What was the label on Lila's chair?

Solution 15

Scalable method:

Figure out the problem for a small number of chairs. The number of chairs must be even for a chair to have an opposite. 4 chairs will do. With 4 chairs, 3 is opposite from 1 and 2 is opposite from 4. So the half of the number of chairs is opposite the total.

You can infer the following rule for even numbers of chairs: to find the number opposite a small number, add half of the total number of chairs to it. Similarly, to find the number opposite a large number, subtract half of the total number of chairs from it. Since Dina sits at a number larger than 9, we have to use subtraction. Lila sits on chair number $14 - 9 = 5$.

Brute force method:

This table shows which chairs are opposite:

1	2	3	4	5	6	7	8	9
10	11	12	13	14	15	16	17	18

From the table, we see that Lila was seated at number 5. This method, however, is much more difficult to apply if there are many more seats.

Exercise 16

15 scouts eat 1050 bagels in 35 days. How many scouts eat 360 bagels in 18 days?

Solution 16

Use the method of reduction to unity to find out how many bagels one scout eats in one day (line 3 in the following table).

scouts	days	bagels
15	35	1050
1	35	70
1	1	2
1	18	36
10	18	360

SOLUTIONS TO PRACTICE TWO

Exercise 1

Lila chose an integer number. Then she told Dina: "If I add half of my number to a third of it, I get 500. Can you guess my number?" What was Dina's reply?

Solution 1

The number Lila chose must be a multiple of 2 as well as a multiple of 3. Therefore, it is a multiple of 6.

Let us divide this number into six equal parts. Three of these parts represent half the number and two of the parts represent a third of the number. These five equal parts total 500. This means each part is equal to 100 and that the number Lila chose was 600.

Exercise 2

The sum of two numbers is 791. The sum of the triple of one of the numbers and the other number is 875. What is the positive difference of the numbers?

Solution 2

From the diagram we see that, by subtracting 791 from 875, we obtain the double of one of the numbers: 84. Therefore, one of the numbers is equal to 42. The other number is equal to: $791 - 42 = 749$.

The positive difference of the two numbers is $749 - 42 = 707$.

Exercise 3

Stephan, the tennis coach, has 32 tennis balls in a bag. Of these, there are four times as many Dunlap balls as Wilson balls. Which of the following statements must be true?

(A) There are 8 Wilson balls.

(B) Stephan must have balls of at least one more brand in the bag.

(C) If you take out a ball without looking, it will be a Dunlap.

(D) Wilson brand is not Stephan's favorite.

(E) The bag is full of balls.

Solution 3

Since there are four times as many Dunlap balls as Wilson balls, the total number of Dunlap and Wilson balls must be a multiple of 5. Since 32 is not a multiple of 5, Stephan must have balls of at least one other brand in the bag.

There cannot be 8 Wilson balls since the total number of balls would then be 40 or more.

The correct answer choice is (B).

Exercise 4

Dina, Lila, and Amira went to pick apricots. Dina picked as many as Lila and Amira combined. Lila picked one fifth the amount Amira picked. Which of the following could be the total number of apricots they picked?

Solution 4

Amira picked five times more fruit than Lila. Together, Amira and Lila picked six times more fruit than Lila. Dina also picked six times fruit more than Lila. In total, the three girls picked 12 times as many apricots as Lila picked by herself.

The total number of apricots has to be a multiple of 12.

The only even numbers among the answer choices are 154 and 132. Of these two, only 132 is a multiple of 12.

Exercise 5

Lila said to Dina: "Let's play with your mini-monsters! You have 94, right?" Dina replied: "I don't have 94 mini-monsters. If I had one third of what I have plus another sixth of what I have, plus the double of what I have, and if I included the 4 farm animals, then I would have 94 toys." How many mini-monsters did Dina actually have?

Solution 5

Since Dina talks about a sixth of her number of mini-monsters as a possible number of mini-monsters, then a sixth must be a whole number. This means that Dina has a number of monsters that is a multiple of 6. Let us call this number $6w$.

A sixth of $6w$ would be w and a third of $6w$ would be $2w$.

In the total of 94 toys, 4 farm animals are included. Thus, there would be 90 mini-monsters.

The total of 90 mini-monsters would be achieved like this:

$$2w + w + 12w = 90$$

which is equivalent to:

$$15w = 90$$

and therefore $w = 6$.

Dina had only 36 mini-monsters.

Exercise 6

Arbax, the Dalmatian, is organizing his bone collection. He finds that half of the bones are way too long and breaks them in half. Then, he finds that 36 of the newly broken bones are still too long and breaks those in half too. In the end, he has 96 pieces to store. How many bones did he start with?

Solution 6

Since Arbax can divide the number of bones in half, he must have had an even number of bones to start with. Let us say that Arbax had $2b$ bones initially. Half of this number is b. Arbax will break b bones in half and keep b bones intact. Let us count the b intact bones towards the total number of pieces Arbax has to store.

b bones will be cut in half and $2b$ bones will result. Of these, $2b - 36$ will remain intact and 36 will be broken in half, resulting in 72 pieces. Let us count the $2b - 36$ and the 72 towards the total number of pieces.

Now, the total number of pieces is:

$$b + 2b - 36 + 72$$

which we know is equal to 96. Solving the equation $3b + 36 = 96$, we find that $b = 20$. Therefore, Arbax must have had 40 bones initially.

Exercise 7

Max, the baker, has been selling bags of 10 pretzels for 5 dollars a piece all day. Now it is evening and Max wants to get rid of as much of his merchandise as he can before it becomes day old. He decides to keep the price of each bag the same but add some more pretzels to each bag so that the final price is 4 dollars per dozen. How many pretzels did Max add to each bag?

Solution 7

If Max sells 10 pretzels for 5 dollars, then 2 pretzels cost one dollar.

After some pretzels were added to the bag, 12 pretzels cost 4 dollars: Max is now selling 3 pretzels for one dollar.

Since the bag still costs 5 dollars, there must be $3 \times 5 = 15$ pretzels in each bag. Max added 5 pretzels to each bag.

Exercise 8

Stephan is moving to another tennis club and is now weighing his boxes to get a moving quote. He has two identical boxes: one with 55 tennis balls, and the other with 15 tennis balls. The box with 55 balls weighs 3 kg and the box with 15 balls weighs 1 kilogram. How much does an empty box weigh?

Solution 8

The empty box will weigh less than 1 kg. To handle weights that are less than 1 kg more comfortably, it is a good idea to convert the weights to grams: 3 kg = 3000 g and 1 kg = 1000 g.

If you subtract the weight of the lighter box from the weight of the heavier box, you will get the weight of the additional 40 balls: 2000 grams. Therefore, each tennis ball weighs 50 grams.

15 balls weigh $15 \times 50 = 750$ grams.

An empty box must weigh: $1000 - 750 = 250$ grams.

Exercise 9

Divide the number 29 into three integer parts so that, if we add 1 to the first part, 2 to the second part, and 4 to the third part, you get consecutive numbers.

Solution 9

The three integer parts must add up to 29. The three consecutive numbers must add up to $29 + 1 + 2 + 4 = 36$. Three consecutive numbers that add up to 36 are: 11, 12, and 13.

The parts are $10, 10$, and 9:

$$
\begin{aligned}
11 - 1 &= 10 \\
12 - 2 &= 10 \\
13 - 4 &= 9
\end{aligned}
$$

Exercise 10

Divide the number 72 into four integer parts so that if we add 5 to the first part, subtract 5 from the second part, multiply the third part by 5, and divide the fourth part by 5, all the results are the same.

Solution 10

Denote the parts as $a, b, c,$ and d. First of all, we know that these parts add up to 72:

$$a + b + c + d = 72$$

Also, we know that:

$$a + 5 = b - 5 = 5 \times c = d \div 5$$

If $c \times 5 = d \div 5$, then $d = 25 \times c$.

Also, $b = 5 \times c + 5$ and $a = 5 \times c - 5$.

Add $a, b, c,$ and d together:

$$5 \times c + 5 + 5 \times c - 5 + c + 25 \times c = 72$$

$$36 \times c = 72$$

Therefore, $c = 2$. From this value, we can build all the other parts:

$$d = 50$$

$$b = 15$$

$$a = 5$$

which together add up to 72.

Exercise 11

Amira's teacher told her to multiply a number by 7 and subtract 5. Amira absent-mindedly multiplied the number by 5 and added 7. However, Amira's answer was correct! What number did the teacher give her?

Solution 11

Let us denote the initial number by N. Then, the correct operations would be:

$$7 \times N - 5$$

The operations Amira performs are:

$$5 \times N + 7$$

The result is the same in both cases:

$$7 \times N - 5 = 5 \times N + 7$$

$$2 \times N = 12$$

Therefore, $N = 6$.

Exercise 12

In a game of Ask-Me-NOT, a player receives 1 point for each correct answer and loses 1 point for each incorrect answer. Stephan was invited to such a game and, overall, was able to give 5 correct answers. Nevertheless, he ended the game with a score of -12. How many questions did he answer in total?

Solution 12

Stephan earned 5 points by answering 5 questions correctly. He lost 5 points by answering 5 questions incorrectly. He got -12 points by answering another 12 questions incorrectly. Therefore, there were 22 questions in total.

Exercise 13

Jared works at a pet store. His job is to sort crickets and worms into lunch sized portions for the lizards and birds that live in the store. He puts either 5 crickets or 7 worms in each box. If he has a total of 46 worms and crickets, how many boxes will he fill?

Solution 13

The total number of worms must be a multiple of 7 and the total number of crickets must be a multiple of 5.

Multiples of 5 end in either 5 or 0.

If the total is 46, then the number of worms must end in either 1 or 6. The only multiple of 7 that ends in 1 and is smaller than 46 is 21. All multiples of 7 that end in 6 are all larger than 46.

Therefore, there must be 3 boxes of worms: $3 \times 7 = 21$ and 5 boxes of crickets: $5 \times 5 = 25$. Jared filled 8 boxes in total.

Exercise 14

In Dina's chicken coop, there are 4 red chickens for every 5 white chickens. The total number of chickens could be:

(A) 61
(B) 63
(C) 75
(D) 100
(E) 101

Solution 14

The chickens in Dina's coop can arranged in groups of 9, with 4 red and 5 white chickens in each group. Therefore, the total number of chickens must be a multiple of 9. Only choice (B) is a multiple of 9.

Exercise 15

There are 14 dolphins playing around Amanda's boat. 12 of them are females and 8 of them are calves. What is the smallest possible number of female calves? What is the largest possible number?

Solution 15

There are 2 male dolphins.

The smallest number of female calves occurs if the 2 males are calves. In this case, there are 6 female calves.

The largest number of female calves occurs if the 2 males are adults. In this case, there are 8 female calves.

There are at least 6 and at most 8 female calves.

Exercise 16

Ali's donkey is so stubborn that Ali has to drag him forward. The donkey will follow Ali for 20 feet, after which he walks backwards for 5 feet and waits for Ali to drag him again. After much struggle, Ali was able to move the donkey 100 feet, from the Fountain to the Market. How many feet did the donkey walk in total?

Solution 16

Solve this problem by making a diagram to retrace the steps of the donkey. It might look like this:

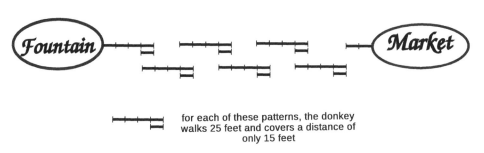

for each of these patterns, the donkey walks 25 feet and covers a distance of only 15 feet

The donkey walks 6 x 25 + 10=160 feet but covers a distance of 100 feet

54

The donkey walked 160 feet in total to cover a distance of 100 feet.

Exercise 17

Dina, Lila, and Amira are cleaning out their toy collections. Lila had 10 dinosaurs more than Dina and 10 dinosaurs fewer than Amira. They sold the dinosaurs they no longer wanted at a yard sale. Dina sold a quarter of her dinos, Lila sold half of hers, and Amira sold 14. They were left with a total of 47 dinos. How many dinos did Dina have in the beginning?

Solution 17

We are going to calculate the total number of dinos in two different ways and set the results to be equal.

Together, Dina and Amira have twice as many dinos as Lila:

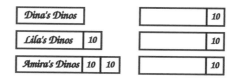

Therefore, the total number of dinos must be a multiple of 3. It is equal to three times Dina's dinos plus 30.

Let us now work backwards and add the 14 dinos Amira sold to the 47 remaining dinos. We get 61 dinos. To these, we add half of Lila's dinos: that is equal to half of Dina's dinos plus 5. The sum has become equal to half of Dina's dinos plus 66.

Now we add one quarter of Dina's dinos back and the sum becomes:

$$\frac{1}{4} + \frac{1}{2} = \frac{3}{4}$$

three quarters of Dina's dinos plus 66.

Three times Dina's dinos plus 30 equals three quarters Dina's dinos plus 66.

$$3 - \frac{3}{4} = \frac{9}{4}$$

Nine fourths of Dina's dinos equal 36. Dina must have 16 dinos, Lila must have 26, and Amira must have 36 dinos.

Exercise 18

Arbax sorted his bones in piles of 6 bones each. After making the piles, he noticed that the number of bones left over was equal to half the number of piles. The total number of bones cannot be:

(**A**) 13

(**B**) 14

(**C**) 26

(**D**) 65

(**E**) 52

Solution 18

If the number of bones left over is equal to half the number of piles, then the number of piles must be even. Also, the number of leftover bones must be smaller than 6, since Arbax was not able to make another pile of 6 bones. Possible values for the number of leftover bones are:

$$\{1,\ 2,\ 3,\ 4,\ 5\}$$

The number of piles can be:

$$\{2,\ 4,\ 6,\ 8,\ 10\}$$

The total number of bones can be:

$$6 \times 2 + 1 = 13$$
$$6 \times 4 + 2 = 26$$
$$6 \times 6 + 3 = 39$$
$$6 \times 8 + 4 = 52$$
$$6 \times 10 + 5 = 65$$

Of the answer choices, only 14 is not in the list of possible numbers.

Exercise 19

Max baked 38 baguettes and put them on display to the left and to the right of his sales counter. Lila comes in with some friends and they each make an observation about the number of baguettes on the left and the number of baguettes on the right:

Solution 19

Answer (A) - there is an equal number of baguettes on both sides - is possible because 38 is even.

Answer (B) - there are three times more baguettes on the left - is not possible because 38 is not a multiple of 4.

Answer (C) - there are six baguettes more on the right - is possible because $38 - 6 = 32$ is even.

Answer (D) - there are four baguettes fewer on the left - is possible because $38 - 4 = 34$ is even.

Answer (E) - there are ten baguettes more on the right - is possible because $38 - 10 - 28$ is even.

Exercise 20

At the Lucky Stars merry-go-round, Lila passes through a tunnel, by a mirror, and over a bottomless pit. She collects 3 points at the tunnel, 4 points at the mirror, and 5 points at the bottomless pit. During the ride, she collected a total of 100 points. How many times did she pass by the mirror?

Solution 20

After each complete turn, Lila has collected a total of 12 points $(3 + 4 + 5 = 12)$.

Dividing 100 into 12 gives a quotient of 8 and a remainder of 4. This means Lila completed 8 rounds and then collected 4 more points. This is only possible if she passed by the mirror one more time. Therefore, Lila passed by the mirror 9 times.

Exercise 21

Arbax thinks to himself: "If I had 8 bones more than I have right now, then I could share them equally with Lynda and still have one bone more than I have now." How many bones does Arbax have?

Solution 21

If the bones are shared equally with Lynda, then Lynda would also have as many bones as Arbax has now plus one more. Together, Arbax and Lynda would have twice as many bones as Arbax has now plus two more.

This means that the extra 8 bones account for the number of bones Arbax has now plus two more bones. Arbax must have 6 bones!

Exercise 22

Max is closing his bakery at the end of the day. He thinks to himself: "If I had sold three times as many baguettes as focaccias or if I had sold twice as many foccacias as baguettes, I would have sold 50 breads more than I did." How many breads did Max sell that day?

Solution 22

If Max had sold three times as many baguettes as focaccias, then the total number of breads sold would have been equal to 4 times the number of foccacias.

If Max had sold two times as many foccacias as baguettes, then the total number of breads sold would have been a equal to 3 times the number of baguettes.

Max states that these two hypothetical situations would have resulted in the same number of loaves sold: three times the number of baguettes is equal to four times the number of foccacias.

Therefore, the hypothetical total is a multiple of 12 that is larger than 50.

$$\boxed{F} + \boxed{F} + \boxed{F} + \boxed{F} \;=\; \boxed{B} + \boxed{B} + \boxed{B}$$

$$\boxed{F} + \boxed{B} \;=\; \boxed{B} + \boxed{B} + \boxed{B} - 50$$

$$\boxed{F} \;=\; \boxed{B} + \boxed{B} - 50$$

$$\boxed{B} + \boxed{B} - 50 + \boxed{B} + \boxed{B} - 50 + \boxed{B} + \boxed{B} - 50 + \boxed{B} + \boxed{B} - 50 = \boxed{B} + \boxed{B} + \boxed{B}$$

Compare the left hand side and the right hand side of the last row in the figure. The comparison shows that 5 times the number of baguettes equals 200. Therefore, there are 40 baguettes.

If there are 40 baguettes, there must be 30 focaccias. Max sold a total of 70 loaves of bread.

Exercise 23

Stephan, the tennis coach, is visiting a new client. This client lives on the same street as another client and Stephan has forgotten which one of them lives at house number 38 and which one at 78. He calls the new client and explains his problem. It is, however, April 1st and Stephan gets the following riddle instead of direct help: "If you multiply my house number by 15 and multiply the other house number by 14, the sum of the two results will be a multiple of three." Help Stephan quickly!

Solution 23

38 is not a multiple of three, but 78 is.

Also, 15 is a multiple of three, but 14 is not.

To obtain a sum that is a multiple of three, if one of the terms is a multiple of three, the other must also be a multiple of three.

Since one term is a house number multiplied by 15, a multiple of three, the other term must be 14 times the house number that is a multiple of three. Therefore, the number that 14 is multiplied by must be 78.

The new client's house number is 38.

Exercise 24

Which of the following is the correct answer to the question: "How many numbers from 0 to 40 are multiples of 7?"

(**A**) {7, 14, 21, 28, 35}

(**B**) 5 numbers

(**C**) 6 numbers

(**D**) only 1 and 7 since 7 is prime

Solution 24

Answer (A) is incorrect because the question asks us to find a single number: the number of multiples.

Answer (B) is incorrect because the number of multiples is 6, since 0 is a multiple of any positive integer, including 7.

Answer (C) is correct.

Answer (D) is incorrect because we are asked to find the multiples of 7, not its divisors.

SOLUTIONS TO PRACTICE THREE

Exercise 1

In a bag of coins, there are quarters, dimes, and nickels. Nyame asks Anansi to exchange as many one dollar bills as he can into quarters. Then, Nyame asks Turtle to exchange as many one dollar bills as he can into dimes. Finally, Nyame asks Firefly to exchange as many one dollar bills as he can into nickels. What is the largest possible dollar amount of the coins left in the bag?

Solution 1

The number of quarters Anansi can take out of the bag is a multiple of four. There can be as many as 3 quarters left in the bag.

The number of dimes Turtle can take out of the bag is a multiple of ten. There can be as many as 9 dimes left in the bag.

The number of nickels Firefly can take out of the bag is a multiple of 20. There can be as many as 19 nickels left in the bag.

The total amount of money left in the bag could be as much as:

$$3 \times 25 + 9 \times 10 + 19 \times 5 = 260 \text{ cents}$$

which is equivalent to 2 dollars and 60 cents.

Exercise 2

If Amira had one doll more than she has, she could arrange her dolls in groups of five exactly. If Dina had two dolls more than she has, she could arrange her dolls in groups of five exactly. If Lila had three dolls less than she has, she could arrange them in groups of five exactly. If the three girls put all their dolls together, would it be possible to arrange them in groups of five without any toys left over?

Solution 2

Let us arrange the dolls each girl has in groups of five dolls. Amira needs one more doll to make a group of five. After making the groups, she has 4 dolls left over. Both Lila and Dina must have 3 dolls left over after making groups of five.

Now, all the dolls the girls have are arranged in groups of five, except for the left over dolls. How many dolls are there left over? $4+3+3 = 10$. Since 10 dolls are left over, the girls can make two additional groups of five dolls.

Exercise 3

Lila was making loot bags for her birthday. She had 8 loot bags and wanted to put chocolates from a larger box in each of them. She placed one chocolate in each bag and continued to do so until each bag contained 13 chocolates. When she had finished, there were some chocolates left in the larger box, but not enough to place one more in each loot bag. At the start, the large box may have contained:

(A) 100 chocolates.

(B) 108 chocolates.

(C) 112 chocolates.

(D) 134 chocolates.

Solution 3

Note that:

$$\text{Dividend} = \text{Divisor} \times \text{Quotient} + \text{Remainder}$$

where the remainder is always smaller than the divisor.

In the problem, since Lila has 8 loot bags, the remainder can only range from 1 to 7. The total number of chocolates must be larger than or equal to:

$$8 \times 13 + 1 = 105$$

and smaller than or equal to:

$$8 \times 13 + 7 = 111$$

The only answer choice in the possible range is 108.

Exercise 4

Dina and Lila divided 122 by 17. Dina got a quotient of 7 and a remainder of 3. Lila got a quotient of 6 and a remainder of 20. Which of the following is true? (check all that apply)

Solution 4

The remainder is always smaller than the divisor - otherwise, the quotient can be increased. Since the divisor is 17, the remainder can only be a number from 0 to 16. Therefore, Lila cannot be correct. Dina got the correct answer.

Exercise 5

Max, the baker, bought several 8 lbs bags of flour. He purchased 136 lbs of flour in total and used a quarter of this quantity the next day. How many bags were left unopened?

Solution 5

First, find out first how many bags there are:

$$136 \div 8 = 17$$

There are 17 bags.

Now, let us find out how the quantity Max used compares to this number. Find out how many pounds of flour Max used:

$$136 \div 4 = 34$$

Max must open five bags so that he has enough flour. Therefore, 12 bags remain unopened.

Exercise 6

Stephan, the tennis coach, ordered some tennis balls. He received 10 boxes, each containing 12 cylinders of 3 balls, and some boxes of loose balls, each containing 12 balls. If he received 408 balls in total, how many boxes of loose balls was he sent?

Solution 6

First, figure out how many balls there are that are packed in cylinders:

$$10 \times 12 \times 3 = 360$$

Then, figure out how many loose balls there are:

$$408 - 360 = 48$$

Now find out how many boxes were filled with loose tennis balls:

$$48 \div 12 = 4$$

Exercise 7

Dina had 12 dinosaurs, Lila had 20 dinosaurs, and Amira had more than 10 dinosaurs. Dina, Lila, and Amira counted all their dinosaurs and found out that they could divide them equally. What is the smallest number of dinosaurs Amira could have had?

Solution 7

Since it was possible to divide the dinosaurs equally, the total number of dinosaurs must have been a multiple of 3. Dina and Lila together have $12 + 20 = 32$ dinosaurs. If we divide 32 by 3, we get a remainder of 2 dinosaurs. Amira must have a number of dinosaurs that leaves a remainder of 1 when divided by 3, so that the one dinosaur together with the 2 dinosaurs left over from Dina and Lila can form another group of 3.

Therefore, we must find the smallest number greater than 10 that gives a remainder of 1 when divided by 3. That number is 13:

$$13 = 4 \times 3 + 1$$

13 is the smallest number of dinosaurs Amira could have had. In this case, there were $12 + 20 + 13 = 45$ dinosaurs in total and each girl would have 15 dinosaurs if they divided them equally.

Exercise 8

Arbax has dug up his bone collection. He made piles of 5 bones each and gave any leftover bones to Lynda. Then he made piles of 4 bones each and again gave any leftover bones to Lynda. Lynda cannot have:

(A) 4 bones

(B) 5 bones

(C) 6 bones

(D) 7 bones

(E) 8 bones

Solution 8

The division by 5 can have a remainder from 0 to 4, while the division by 4 can have a remainder from 0 to 3. The smallest number of bones Lynda could have is 0 and the largest is 7 - any number of bones within these boundaries is possible.

Lynda cannot have 8 bones. The answer is (E).

Exercise 9

Alfonso, the grocer, is filling some gift baskets with apples and mangoes. He does not want to count all the fruit but he notices that, if he tries to put an equal number of apples in each basket, there are 4 apples in each basket and 8 apples left over. If he tries to put an equal number of mangoes in each basket, there are 5 mangoes in each basket and 5 mangoes left over. If there are 7 more mangoes than apples, can you find out how many baskets he has?

Solution 9

If the number of apples left over is 8, then there must be more than 8 baskets (the remainder is always smaller than the divisor!).

If Alfonso had 3 mangoes more than he has now, there would be 8 of each type of fruit left over. Also, there would be 10 more mangoes than apples. In this case, since the leftover is the same, the difference between the number of mangoes and the number of apples must be equal to the number of baskets. This is because, in each basket, the number of mangoes exceeds the number of apples by one. There are 10 baskets.

For those who are comfortable with expressions, denote the number of apples by A, the number of mangoes by M, and the number of baskets by B. Write the integer division for apples:

$$A = 4B + 8$$

Write the integer division for mangoes:

$$M = 5B + 5$$

Add 3 mangoes:

$$M + 3 = 5B + 8$$

Subtract the apples from the mangoes:

$$M + 3 - A = 5B - 4B$$

Use the fact that $M - A = 7$:

$$
\begin{aligned}
7 + 3 &= B \\
B &= 10
\end{aligned}
$$

Exercise 10

Dina and Lila have set up 61 decorations for Hallowe'en. After they finished, Lila said to Dina: "I put up 5 more decorations than a third of the number of decorations you put up." How many decorations did Dina put up?

Solution 10

The number of decorations Dina put up must be a multiple of 3. Dina put up three thirds and Lila put up another third plus an extra 5. The total of 61 is formed of the 4 thirds plus an extra 5 decorations. Subtract the 5 extra decorations from the total $(61 - 5 = 56)$ and divide the result by 4 to get the number of objects that represent one third:

$$56 \div 4 = 14$$

Each third has 14 objects. Therefore, Dina put up $14 \times 3 = 42$ decorations.

Exercise 11

Max, the baker, has just finished baking his daily supply of bread. He notices there are 4 baguettes for each boule and 6 rolls for each baguette. The total number of baguettes, boules, and rolls is 290. How many are there of each?

Solution 11

There are 24 rolls for each boule. Therefore, there are $4 + 24 = 28$ baguettes and rolls for each boule. As a result, it is possible to arrange the bread varieties in 10 groups of 29. There must be 10 boules, 40 baguettes, and 240 rolls.

Exercise 12

Lila, Dina, and Amira helped Stephan harvest the pomegranates in his tree. Each girl put the pomegranates she harvested in boxes of 8 and tossed the leftover fruit in a basket. At the end of the harvest, what was the largest number of boxes that could be filled from the contents of the basket?

Solution 12

The amount of leftover fruit that each girl may have ranges from 0 to 7. Since there are three girls, the largest number of fruits that could end up in the basket is $3 \times 7 = 21$. Divide 21 by 8 to find the number of boxes that could be filled:

$$21 = 8 \times 2 + 5$$

2 more boxes could be filled.

SOLUTIONS TO PRACTICE FOUR

Exercise 1

Four clocks show the following times:

(A) 4:00 PM

(B) 4:30 PM

(C) 4:45 PM

(D) 4:50 PM

One of them stopped 15 minutes before Dina left for school. Another stopped one third of an hour after Dina arrived at school. The other two clocks are broken and show random times. If it takes Dina 10 minutes to walk from home to school, which clocks show random times?

Solution 1

First, convert the third of an hour into minutes. A third of an hour equals 20 minutes. Then, make a timeline to visualize the data:

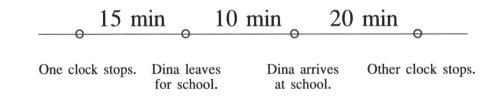

15 min	10 min	20 min	
One clock stops.	Dina leaves for school.	Dina arrives at school.	Other clock stops.

We have to look for two clocks that are $15 + 10 + 20 = 45$ minutes apart. Therefore, (A) and (C) are the stopped clocks, while (B) and (D) show random times.

Exercise 2

Lila left for her singing class at 4:20 PM and returned at 7:10 PM. For how many minutes was Lila away from home?

Solution 2

There are 40 minutes between 4:20 PM and 5:00 PM, 2 hours between 5:00 PM and 7:00 PM, and 10 minutes between 7:00 PM and 7:10 PM. Lila was away from home for $40 + 120 + 10 = 170$ minutes.

Exercise 3

Dina's mother is four times older than Dina. Dina's age is 35 years less than her father's. Dina's father was 5 years older than Dina's mother when Dina was born. How old is Dina now?

Solution 3

Dina is 10 years old. Use the following diagram to make sense of the information in the problem:

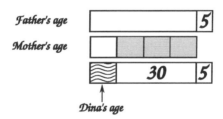

From the figure, we see that the shaded rectangles must total 30. Therefore, each shaded rectangle must represent 10. Dina is 10 years old, her mother is 40, and her father is 45.

Exercise 4

Amalia, the dance instructor, has started her summer classes. On weekdays, she teaches from 9:00 AM to 11:30 AM and from 1:00 PM to 5:00 PM. On weekends, she teaches from 1:00 pm to 4:00 pm. If all her classes are the same length, each class cannot be:

(A) 10 minutes long.

(B) 15 minutes long.

(C) 20 minutes long.

(D) 12 minutes long.

Solution 4

Amalia works for:

$$2 \times 60 + 30 = 150 \text{ minutes on weekday mornings,}$$

$$4 \times 60 = 240 \text{ minutes on weekday afternoons,}$$

$$\text{and } 3 \times 60 = 180 \text{ minutes on Saturday and Sunday.}$$

Of these intervals, the 150 minute interval one cannot be divided in class sessions of 12 minutes each. Therefore, classes cannot be 12 minutes long.

Exercise 5

Dina and Lila each got an hourglass for their birthday. The two hourglasses are not the same size. Lila has to turn her hourglass 4 times in 5 hours and Dina has to turn her hourglass 6 times in 5 hours. What is the difference, in minutes, between the intervals of time measured by the two hourglasses?

Solution 5

First, see how many minutes 5 hours represent: $5 \times 60 = 300$ minutes.

Lila turns her hourglass every 75 minutes:

$$300 \div 4 = 75$$

Dina turns her hourglass every 50 minutes:

$$300 \div 6 = 50$$

The two hourglasses measure time intervals that differ by 25 minutes.

Exercise 6

Express the following military times as AM or PM.

Solution 6

1. 06:00 is 6 AM
2. 12:30 is 12:30 PM
3. 23:45 is 11:45 PM
4. 19:12 is 7:12 PM
5. 02:30 is 2:30 AM
6. 14:15 is 2:15 PM
7. 11:30 is 11:30 AM

Solution 7

4:45. The hands of the clock have been switched!

Exercise 8

Amira broke her piggy bank to buy herself a movie ticket. After she counted her savings, she exclaimed: "If I had 20 dimes more, 5 nickels less, and 5 pennies less, then I would have six times the amount I have now!" How much money, in dollars, did Amira find in the piggy bank?

Solution 8

20 dimes are 200 cents, 5 nickels are 75 cents, and 5 pennies are 5 cents. The difference of $200 - 75 - 5 = 200 - 80 = 120$ cents represents 5 times the amount Lila has now. Therefore, Lila has 24 cents.

Exercise 9

Amira's mother was in her garden, tying tomato plants to wooden stakes: it took her 10 minutes to stake each plant and she placed the plants 2 feet apart. If she started at one end of the row at 8:00 AM, how far from the first plant was the tomato plant she finished tying up at 9:00 AM?

Solution 9

Amira's mother finishes staking the first tomato at 8:10 AM. At 9:00 AM, she finishes staking the sixth plant down the row. There are 5 gaps between 6 plants and each gap is 2 feet long. The first plant and the sixth plant are $5 \times 2 = 10$ feet apart.

Exercise 10

Write the next term in the sequence:

022814, 033115, 043016, 053117, ...

Solution 10

The terms are dates: February 28, 2014, March 31, 2015, April 30, 2016, May 31, 2017. We can follow this with June 30, 2018: **063018**.

Exercise 11

Dina rides on horseback from one end of a trail while Lila rides on horseback from the other end of the trail. Dina rides mostly uphill, so her horse moves at an average of 2 miles per hour. Lila rides mostly downhill. Her horse averages 3 miles per hour. Lila and Dina meet after 90 minutes of riding. How long is the trail?

Solution 11

90 minutes are equal to one hour and a half. Dina rode 3 miles while Lila rode 4.5 miles. The length of the trail is 7.5 miles.

Exercise 12

Dina and Lila went cross country skiing. They left from different ends of the same trail and decided to meet at a lodge which was midway. Dina started 3 miles from the lodge and Lila started 5 miles from the lodge. If they both skied at the same average speed of 2 miles per hour and started at the same time, how long did Dina have to wait for Lila at the lodge?

Solution 12

By the time Dina reaches the lodge, Lila still has 2 more miles to go. Since Lila's average speed is 2 miles per hour, Dina will have to wait one hour.

Exercise 13

Dina's mother can clean the floor of the house in 3 hours. Dina's father can do the same job in 2 hours. How long does it take them to clean the floor if they work at the same time?

Solution 13

In 1 hour, Dina's mother can clean one third of the floor while Dina's father can clean one half of the floor. In one hour, both parents can clean:
$$\frac{1}{3} + \frac{1}{2} = \frac{5}{6}$$
five sixths of the floor. They can clean one sixth of the floor in one fifth of an hour. One fifth of an hour is equal to 12 minutes. It takes Dina's parents 1 hour and 12 minutes to clean the floor.

Exercise 14

Dina's mother has purchased a robot that can clean the floor of the house in 2 hours. Dina's father did not know about this, so he also purchased the same model of robot. How long does it take the two robots to clean the floor if they work together?

Solution 14

Unlike people, the cleaning robots clean in a random manner and are not capable of working as a team. Set to work simultaneously without supervision, they will clean the same surfaces and it will still take them 2 hours to finish the job.

Exercise 15

Max left for Carapatas and drove at an average speed of 80 miles per hour. After 2 hours, Max was stopped and given a traffic ticket for speeding. Humbled, he continued at an average speed of 70 miles per hour. The distance from Max's home to Carapatas is 440 miles. How long did it take Max to get from his home to Carapatas if his encounter with the traffic police took 15 minutes?

Solution 15

Max traveled $80 \times 2 = 160$ miles in the first two hours. The remaining distance was $440 - 160 = 280$ miles. At a speed of 70 miles per hour, Max traveled this distance in 4 hours. Including the traffic stop, his journey took 6 hours and 15 minutes.

Exercise 16

Amira bikes on the beach trail from the parking lot to the kayak store. Dina starts biking from the parking lot 10 minutes after Amira started and arrives at the kayak store 10 minutes before Amira. If Dina rides twice as fast as Amira, how long did Amira ride?

Solution 16

Dina's total travel time was 20 minutes shorter than Amira's. Since she rode twice as fast as Amira, Amira must have traveled for 40 minutes.

Solution 17

minutes	30	20	45	12	6	18	24	48	25
fraction of one hour	$\frac{1}{2}$	$\frac{1}{3}$	$\frac{3}{4}$	$\frac{1}{5}$	$\frac{1}{10}$	$\frac{3}{10}$	$\frac{2}{5}$	$\frac{4}{5}$	$\frac{5}{12}$

Exercise 18

Amalia, the dance instructor, has been invited to dinner by Alfonso, the grocer. Amalia figured she needed 45 minutes to get from her studio to the restaurant. Not wanting to be late, she left 15 minutes early. There was some roadwork on the way and traffic was reduced to one lane. She arrived at the restaurant at 6:20 PM. Alfonso was already there and exclaimed: "You are only 10 minutes late!" When should Amalia have left to be sure that she arrived on time?

Solution 18

If Amalia was 10 minutes late, their meeting time must have been 6:10 PM. Amalia left at 5:10 PM. She should have left at 5:00 PM.

Exercise 19

As Dina walked on stage for her piano recital, she noticed that the time shown in military notation had the largest digit sum possible. What time did Dina walk on stage?

Solution 19

The time with the largest possible digit sum is 19 : 59.

Exercise 20

Amira's cousin was born on the day in the 1990's when the date (in the form month/day/year) had the largest possible digit sum. What was the exact date of his birth?

Solution 20

In the 1990's, the largest possible digit sum occured on 09/29/1999 (September 29, 1999).

SOLUTIONS TO MISCELLANEOUS PRACTICE

Exercise 1

At a party, there are only groups of three siblings. As it happens, there is no group with only girls or only boys. Moreover, there are more than 2 groups in total and the number of boys equals the number of girls. The total number of girls can be:

(A) any multiple of 3.

(B) only an odd multiple of 3.

(C) only an even multiple of 3.

(D) any multiple of 3 larger than 3.

Solution 1

A group of siblings can only consist of 2 boys and one girl or of 2 girls and one boy. Since there are equal numbers of boys and girls, the groups of siblings can be paired to form groups of 3 boys and 3 girls. There are more than 6 children in total, therefore the possible totals are: 12, 18, 24, etc. The possible numbers of girls form the sequence: 6, 9, 12, etc. The correct answer is (D).

Exercise 2

Alfonso's house is at number 101 on Grapevine Street. There are five more houses from there until the end of the street. How many houses are there on Alfonso's side of the street (House numbers on the street start at 1 and they are odd on one side, even on the other)?

Solution 2

The last house is 111. From 1 to 111 there are 56 odd numbers. Therefore, there are 56 houses on Alfonso's side of the street.

Exercise 3

Stephan, the tennis coach, set up a mock tournament for 4 of his students. Every time a student played another student, Stephan gave 5 dollars to the winner. If each student played all the other students, how much money did Stephan pay out in prizes?

Solution 3

If 4 students play each other, $3 + 2 + 1 = 6$ games are played in total. In each game there is one winner. Therefore, Stephan paid out 30 dollars for awards.

Exercise 4

At running, Amira is the worst of the first 10 students in her class as well as the best of the last 10 students in her class. How many students are there in Amira's class?

Solution 4

19. There are 9 students who are faster than Amira and 9 students who are slower than Amira. Therefore, there are $9 + 9 + 1 = 19$ students in the class.

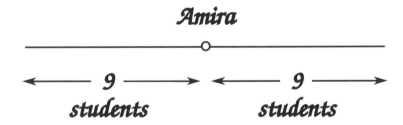

Exercise 5

Julien, the Parisian master baker, only taught his secret French baguette recipe to one of his apprentices: Max. Max did not keep the recipe a secret and taught it to four of his apprentices. These apprentices went on to teach it to four apprentices each, who all taught the recipe to four more apprentices each as well. How many bakers know Julien's recipe now?

Solution 5

86. Max taught it to 4 people. These people taught it to 16 people. The 16 people then taught it to 64 people. In total, $1 + 1 + 4 + 16 + 64 = 86$ people know the recipe. Max and Julien are included in the total, as they should be.

Exercise 6

Alfonso, the grocer, gave Lila a job over the summer: to canvas the neighborhood and take orders for deliveries of fresh vegetables. Lila had a sales quota of 60 dollars per day, but her neighbors trusted Alfonso so much that she managed to take orders of 80 dollars each day. At this pace, she fulfilled the total sales quota 5 days before her job officially ended. For how many days was her job supposed to last?

Solution 6

Every day, Lila's sales exceed her target by 20 dollars. Every 3 days, Lila's extra sales make up for a whole day of work. In order to get ahead by 5 days, Lila worked for $3 \times 5 = 15$ days. Her original work assignment was supposed to last 20 days.

Exercise 7

Dina takes Russian language lessons each Monday, Wednesday, and Friday. What is the largest possible number of lessons Dina could take in any one month?

Solution 7

The largest possible number of lessons can be taken in a 31-day month that starts on a Monday or a Wednesday. In that case, there would be

14 lesson days. A possible example is shown in the figure:

Su	M	Tu	W	Th	F	Sa
	1	2	3	4	5	6
7	8	9	10	11	12	13
14	15	16	17	18	19	20
21	22	23	24	25	26	27
28	29	30	31			

Exercise 8

Max, the baker, is placing fresh loaves of bread on the display shelf of his bakery. On the topmost shelf there are some large loaves of bread. On the middle shelf, there are three times more loaves than on the top shelf. On the bottom shelf, there are three times more loaves than on the middle shelf. The loaves on display are only one third of the number of loaves that Max baked that morning. In total, Max baked 585 loaves. How many loaves are there on the middle shelf?

Solution 8

Picture the number of loaves on the top shelf as a rectangle. Then, figure how many such rectangles there are for each type of bread:

☐ *Top shelf*
☐☐☐ *Middle shelf*
☐☐☐☐☐☐☐☐☐ *Bottom shelf*

There are $1 + 3 + 9 = 13$ rectangles. Since this is only a third of the total, then Max must have 39 rectangles worth of loaves. Divide 585 by 39 to see how many loaves there are in each rectangle.

$$585 \div 39 = 15$$

There are 15 loaves in each rectangle. Therefore, there are $15 \times 3 = 45$ loaves on the middle shelf.

Exercise 9

Stephan organized a local tennis tournament. His suppliers sent 4 boxes of tennis balls. In each box, there were 10 bundles and in each bundle there were 12 cylinders of three balls. On the first day of the tournament, Stephan placed one box in each of the two courts. The players in one of the cours used 43 balls, while the players in the other court used 53 balls. At the end of the day, how many tennis balls were still in unopened cylinders?

Solution 9

Compute the total number of cylinders:

$$4 \times 10 \times 12 = 480$$

To use 43 balls, 15 cylinders must be opened. To use 53 balls, 18 cylinders must be opened. Compute the total number of cylinders that were opened:

$$15 + 18 = 33$$

Subtract to find the number of unopened cylinders: $480 - 33 = 447$. Multiply by the number of balls in each cylinder: $447 \times 3 = 1341$.

Exercise 10

Jared works at a pet store and his duties include separating the daily shipment of crickets into boxes. Today, the shipment contained between 110 and 150 crickets, but no one at the store knew exactly how many. Jared tried to put 5 crickets in each box, but one cricket was left over. Then, he tried to put 6 crickets in each box, but once again, one cricket was left over. How many crickets were there in the shipment?

Solution 10

If Jared had one cricket less, the crickets would fit exactly into groups of 5 as well as groups of 6: the crickets would form groups of 30. Therefore, there are at least 31 crickets. Since we know the number of crickets is between 110 and 150, we select a multiple of 30 that is in that range, such as 120. There are 121 crickets.

Exercise 11

Jared has been given an even harder problem to solve. He has to package worms and roaches! If he tries to make packages with one worm and one roach in each, there are 15 roaches left over. If he tries to make packages with 2 roaches and one worm in each, there are 17 worms left over. How many roaches and worms does he have to package?

Solution 11

If Jared packs one roach and one worm to a box, then there is an equal number of roaches and worms in each box. Since 15 roaches are left over, the number of worms must be smaller than the number of roaches by 15.

Since there are 17 worms left over after Jared packs 2 roaches and one worm per box, the number of packed worms is 17 less than the total number of worms. We know that the total number of worms is 15 less than the total number of roaches. Therefore, the number of packed worms is equal to the number of roaches minus 32.

Since there are 2 roaches to a worm in the boxes, the number of roaches is equal to twice the number of packed worms. Therefore, the number of roaches is equal to twice the number of roaches minus 64. There are 64 roaches and $64 - 15 = 49$ worms in total.

Exercise 12

In Amira's class, there are 20 students. Their teacher gives them points for all sorts of wonderful reasons. As it happens, each student has a different number of points. The student with the most points has 100 points. The teacher says: "I have noticed that, if each of you would give one point to every student who have less points than you do, then you would all have the same number of points." How many points does the student with the smallest number of points have?

Solution 12

The student with the most points would give away 19 points but would not receive any. The student with the fewest points would receive 19 points but would not give any away. After the exchange, the two students would have the same number of points. Therefore, the difference between the largest number of points and the smallest number of points is 38. The student with the fewest points has 62 points.

Exercise 13

Dina's mother purchased vegetables from Alfonso's shop. A pound of red peppers cost twice as much as a pound of sweet potatoes. By weight, Dina's mother put twice as large a quantity of sweet potatoes as of red peppers in a bag. Alfonso told her she had to pay 20 dollars but Dina's mother did not have enough money in her purse. Alfonso said: "If you purchase only half of the sweet potatoes, the total will be exactly as much as you have." How much money did she have?

Solution 13

In the amounts Dina's mother selected initially, the potatoes and the peppers cost the same amount. Since Dina's mother has to pay 20 dollars, both the potatoes and the peppers cost 10 dollars. Since she purchased only half the amount of peppers, the total was reduced to 15 dollars. Dina's mother had only 15 dollars in her purse.

Exercise 14

On Monday morning at 8:00 AM, Jared started to clean the 28 fish tanks in the pet store. It took Jared 25 minutes to clean each of the first 8 tanks. After that, he made some improvements to his process and cut 5 minutes from the cleaning time per tank for all of the remaining tanks. If Jared spent 4 hours per day cleaning the tanks, what day of the week was it when he finished?

Solution 14

It will be Wednesday. Jared will use $8 \times 25 = 200$ minutes for the first 8 tanks and $20 \times 20 = 400$ minutes for the remaining 20 tanks. He will work 600 minutes in total, which is equivalent to 10 hours of work. Jared cleans tanks for 4 hours on Monday, 4 hours on Tuesday, and 2 hours on Wednesday.

Exercise 15

In a singles tennis tournament, pairs of opponents are selected at random for each round. The winners qualify for the next round. After four rounds, there were 4 players in the semi-finals. How many players participated in total?

Solution 15

After 3 rounds there were 8 players left in the tournament, after 2 rounds there were 16 players left, and after 1 round there were 32 players left. There were 64 participants in the tournament.

Exercise 16

A basket with 10 potatoes in it weighs 5 lbs. If Alfonso adds another 10 potatoes, it weighs 7 lbs. Alfonso wants to figure out how much the empty basket weighs. Can you help him?

Solution 16

The 10 extra potatoes weigh 2 lbs. If the potatoes are all the same size and weight, then the initial 10 potatoes also weigh 2 lbs. The empty basket weighs 3 lbs.

Exercise 17

In 5 hours, 20 gophers dig 50 feet of gallery.

Solution 17

Apply the method of reduction to unity:

1. How many feet of gallery do 24 gophers dig in 3 hours?

hours	gophers	feet	description
5	20	50	20 gophers dig 50 feet in 5 hours
1	20	10	20 gophers dig 10 feet in 1 hour
1	1	$\dfrac{10}{20} = \dfrac{1}{2}$	1 gopher digs $\dfrac{1}{2}$ a foot in 1 hour
1	24	$\dfrac{24}{2} = 12$	24 gophers dig 12 feet in 1 hour
3	24	$12 \times 3 = 36$	24 gophers dig 36 feet in 3 hours

2. How many hours does it take 12 gophers to dig 60 feet of gallery?

hours	gophers	feet	description
5	20	50	20 gophers dig 50 feet in 5 hours
1	20	10	20 gophers dig 10 feet in 1 hour
1	1	$\frac{10}{20} = \frac{1}{2}$	1 gopher digs $\frac{1}{2}$ a foot in 1 hour
1	12	$\frac{12}{2} = 6$	12 gophers dig 6 feet in 1 hour
10	12	6×10	60 gophers dig 60 feet in 10 hours

3. How many gophers will it take to dig 30 feet of gallery in 2 hours?

hours	gophers	feet	description
5	20	50	20 gophers dig 50 feet in 5 hours
1	20	10	20 gophers dig 10 feet in 1 hour
1	1	$\frac{10}{20} = \frac{1}{2}$	1 gopher digs $\frac{1}{2}$ a foot in 1 hour
2	1	$\frac{2}{2} = 1$	1 gophers digs 1 foot in 2 hours
2	30	30	30 gophers dig 30 feet in 2 hours

Exercise 18

Amira wrote lyrics for a song: "3 hours of homework, 2 hours of piano, are worth 8 hours in the sun. 2 hours of homework, 4 hours of piano, are worth 8 hours in the sun, eiiiiiight hours in the sun!" Now, she is trying to figure out how many hours in the sun one hour of piano is worth. Can you help her?

Solution 18

Multiply the first line of the song by two to find that 6 hours of homework and 4 hours of piano are worth 16 hours in the sun. Compare this with the second line to find that 4 hours of homework are worth 8 hours in the sun. Therefore, one hour of homework is worth 2 hours in the sun and 2 hours of homework are equivalent to 4 hours in the sun. From the second line, you see that 4 hours of piano must also be equal to 4 hours in the sun. One hour of piano is equivalent to one hour in the sun.

Exercise 19

Together, two taps fill an empty cistern with water in 4 hours. One of the taps can fill the cistern in 12 hours. How many hours does it take the other tap to fill the cistern?

Solution 19

The tap that can fill the cistern in 12 hours will fill one third of the cistern in 4 hours. The other tap fills the remaining two thirds of the cistern in 4 hours, one third of the cistern in 2 hours, and the whole cistern in 6 hours.

Exercise 20

A farmer has 4 horses, 2 cows, and 20 sheep. A cow eats three times as much as a sheep and a horse eats four times as much as a sheep. The farmer purchased 693 rations of feed, enough to last them 33 days. (The farmer purchases feed for a month but he likes to have a bit extra just in case.) After one week, however, the farmer took in his neighbor's 10 sheep as his neighbor was going on vacation to Hawaii. How many days earlier than expected did he run out of feed?

Solution 20

In terms of feed, 2 cows are equivalent to 6 sheep and 4 horses are equivalent to 16 sheep. Therefore, the farmer must feed the equivalent of $6 + 16 + 20 = 42$ sheep. Since 693 rations last 33 days, the animals eat 21 rations per day. In the first week, the animals ate $21 \times 7 = 147$ rations and $693 - 147 = 546$ rations were left over.

After his neighbor dropped off more sheep, the farmer had to feed the equivalent of $42 + 10 = 52$ sheep. Since the equivalent of 42 sheep eat 21 rations per day, each sheep eats half a ration per day. Therefore, 52 sheep eat 26 rations per day. The remaining 546 rations will last for $546 \div 26 = 21$ days. The farmer ran out of feed $33 - 21 = 12$ days sooner than expected.

Exercise 21

Each group of four numbers in the sequence follows the same pattern. Which number is hidden by the question mark?

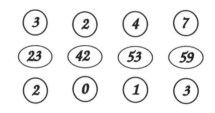

Solution 21

The pattern represents integer division. The number in the rectangle is divided by the number in the top circle and the remainder is placed in the bottom circle.

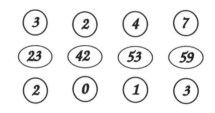

Competitive Mathematics Series for Gifted Students

Practice Counting (ages 7 to 9)
Practice Logic and Observation (ages 7 to 9)
Practice Arithmetic (ages 7 to 9)
Practice Operations (ages 7 to 9)

Practice Word Problems (ages 9 to 11)
Practice Combinatorics (ages 9 to 11)
Practice Arithmetic(ages 9 to 11)
Practice Operations (ages 9 to 11)

Practice Word Problems (ages 11 to 13)
Practice Combinatorics (ages 11 to 13)
Practice Number Theory (ages 11 to 13)
Practice Algebra and Operations (ages 11 to 13)
Practice Geometry (ages 11 to 13)

Practice Word Problems (ages 12 to 15)
Practice Algebra and Operations (ages 12 to 15)
Practice Geometry (ages 12 to 15)
Practice Number Theory (ages 12 to 15)
Practice Combinatorics and Probability (ages 12 to 15)

This is a series of practice books. With the exception of a few reminders, there are no theoretical explanations. For lessons, please see the resources indicated below:

Find a set of free lessons in competitive mathematics at www.mathinee.com. Addressing grades 5 through 11, the *Math Essentials* on www.mathinee.com present important concepts in a clear and concise manner and provide tips on their application. The site also hosts over 400 original problems with full solutions for various levels. Selectors enable the user to sort essentials and problems by test or contest targeted as well as by topic and by the earliest grade level they can be used for.

Online problem solving seminars are available at www.goodsofthemind.com. If you found this booklet useful, you will love the live problem solving seminars.

29684551R00054

Made in the USA
Lexington, KY
03 February 2014